Burt Munro
Presents

SONGBIRDS
Coloring Book

1. Ruby throated hummingbird   Page 3
2. Northern Cardinal   Page 7
3. Baltimore Oriole   Page 9
4. American Robin   Page 10
5. Downy Woodpecker   Page 11
6. American Goldfinch   Page 13
7. Clack Capped Chickadee   Page 15
8. Eastern Bluebird   Page 17
9. Carolina Wren   Page 19
10. Chipping Sparrow   Page 21
11. Yellow Rumped Warbler   Page 23
12. Blue Jay   Page 25
13. Purple Martin   Page 27
14. Mourning Dove   Page 29
15. Killdeer   Page 31
16. Brown Creeper   Page 33
17. House Finch   Page 35
18. White Breasted Nuthatch   Page 37
19. Cedar Waxing   Page 39
20. Tufted Titmouse   Page 41
21. Western Tanager   Page 43
22. Western Meadow hawk   Page 45
23. Rose-breasted Grosbeak   Page 47
24. Northern Mockingbird   Page 49
25. Indigo Bunting   Page 51

Copyright © 2021 Burt Munro
All rights reserved. This book or any portion thereof may not be reproduced or used in any manner whatsoever without the express written permission of the publisher. No reproduction may be made, weatherby photocopying or by other means unless a license has been obtained from the publisher.
First Available, 2021

www.ingramcontent.com/pod-product-compliance
Lightning Source LLC
Chambersburg PA
CBHW082114220526
45472CB00009B/2168